INSECTS

A Matter-of-Fact Book

Written by
RONALD N. ROOD

Illustrated by
CYNTHIA ILIFF KOEHLER
ALVIN KOEHLER

Design, Layout, and Editorial Production by
DONALD D. WOLF and MARGOT L. WOLF

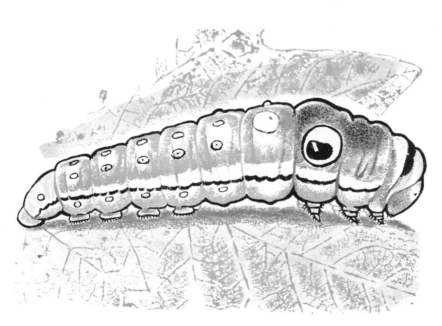

Grosset & Dunlap • Publishers • New York

Some of the material in this book originally appeared in the How and Why Wonder Books of: *Butterflies and Moths; Ants and Bees;* and *Insects*, published by Grosset & Dunlap, Inc.

Library of Congress Catalog Number: 82-80882
ISBN: 0-448-0487-5

Copyright © 1960, 1962, 1963, 1973, 1978, 1982, by Grosset & Dunlap, Inc.
All rights reserved.
Published simultaneously in Canada.
Printed in the United States of America.

Contents

	page		page
Early Insects and Fossils	5	The Insect Parade	40
The Adult Insect	8	1. RELATIVES OF GRASSHOPPERS	42
Life in a Suit of Armor	12	2. THE DRAGONFLIES	42
Insects Helpful to Man	16	3. THE FLIES	43
Eggs by the Thousands	19	4. THE BEETLES	44
Insect Babies—All Shapes and Sizes	26	5. THE TRUE BUGS	45
		6. THE MOTHS AND BUTTERFLIES	46
The Unlucky Caterpillar	30	THE "SOCIAL INSECTS"	49
The "Sleeping Puppet"	34	7. ANTS BEES AND WASPS	51
Jack Frost Arrives	39	8. THE TERMITES	58
		Insects and Plants	60
		Collecting Insects	62

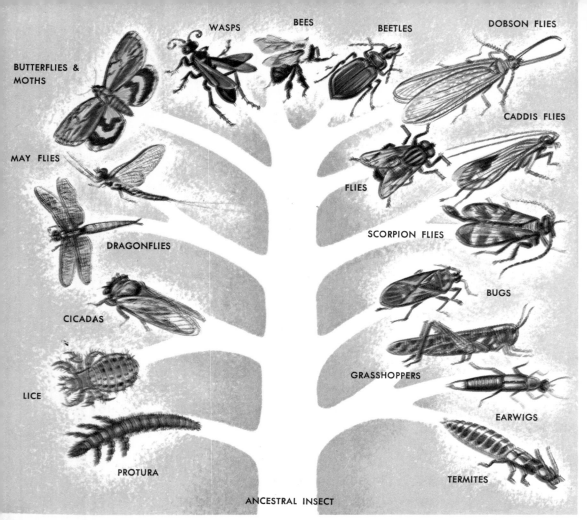

Family tree of the insects.

Primitive reptiles, amphibians, and giant dragonflies were the inhabitants of the coal-age swamps about 250 million years ago.

Early Insects and Fossils

If you could go back in time about one million years, you would come to the days of early human beings. Go back about 100 million years and you would see great dinosaurs. Go back still further and you would see huge insect monsters. Cockroaches as large as saucers would run over the ground. Insects looking like giant dragonflies would sail through the air like model airplanes. Their wings would be more than two feet across—almost seven times larger than many of our common dragonflies today.

These huge insects lived at a time when much of the earth was warm, and food was easy to find. The many plants grew so thickly that they formed great heaps of plant material. This plant material was later compressed and hardened into coal. After the coal-forming period, the land grew cold and dry. Many insects died in the harsh climate. Perhaps they died from other causes, too. Scientists are still trying to find the reasons, like detectives solving a mystery millions of years old.

Many of the smaller kinds of insects survived after the coal-forming period. Their descendants still crawl and fly around today. So you see that insects have been on the earth for a long time. Many scientists believe that they will still be here when all other animal life is gone.

Could there possibly be a giant insect somewhere as large as a human being? Scientists do not think so. Creatures that are very large need some kind of blood system to carry oxygen to all parts of the body. Insects have

Trapped in amber millions of years ago, these well-preserved insects tell the story of the past.

blood, but it doesn't carry oxygen. An insect's breathing pores and air tubes wouldn't provide enough oxygen for an insect the size of a man. Also, since an insect does not have any bones inside its body, a man-sized insect would have to have a heavy jacket of armor for strength. Such armor would make it much too slow and clumsy to survive.

Sometimes one of the insects that lived millions of years ago would land in soft mud or clay, get stuck there, and eventually die. The dead insect would often become completely buried in the mud, which might later turn into rock. When the dead insect within the rock wasted away, it left a natural print, or mold, of its body. Centuries later, if the rock is broken, a picture-outline of the insect may be seen. Such prints and molds are known as fossils.

One of the most interesting insect fossils is the amber fossil. Many kinds of trees, such as pines and spruces, have a sticky material known as resin. You may find it on the bark and trunk. Flies, ants, wasps, and other insects often get tangled in this resin. More of the resin may flow over them, covering them with a clear coating. Later this resin may slowly change and harden, becoming a substance known as amber. If there are insects inside, they will be preserved for millions of years by the hard material.

No one can be sure just when the very first insects lived on the earth. Scientists have found insect fossils that are about 240 million years old. Someday they may find some that are still older.

The more we look at the world of insects, the more interesting it is. We can see amazing creatures such as the tiger beetles, which have long hairs on

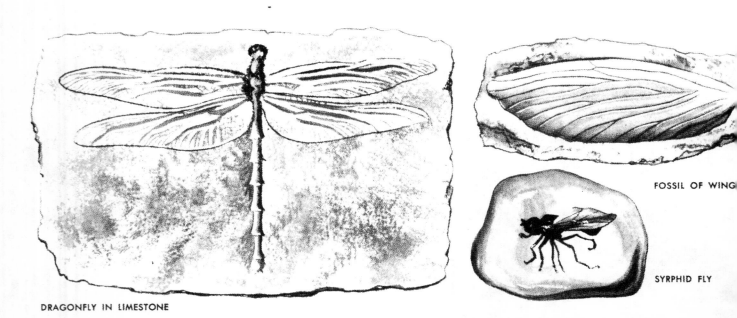

DRAGONFLY IN LIMESTONE

FOSSIL OF WING

SYRPHID FLY

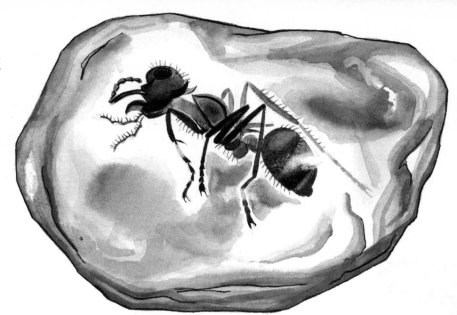

Most animals changed a great deal in the course of millions of years of the earth's history. Many became extinct. But the ants have survived and changed very little, as is apparent from this early specimen. This ant was trapped and preserved in resin millions of years ago.

their feet so they won't sink into the sand dunes where they live. We can watch slug caterpillars that seem to move along like little bulldozers. Gold beetles look as if they were made of pure gold. Tumblebugs roll little balls of material along like children making a snowman. Water pennies look like coins crawling slowly along the bottom of a stream.

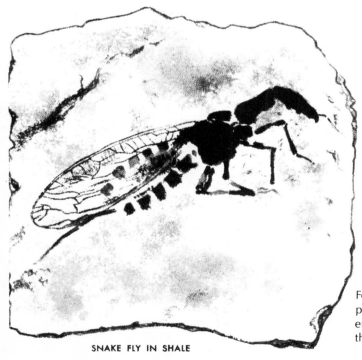

SNAKE FLY IN SHALE

Fossil remains of insects and parts of insects have given scientists useful knowledge about the earth.

The Adult Insect

Is a spider an insect? How about a centipede or scorpion? Are crabs and lobsters really big insects that live in the water? Maybe you have seen a tick on a dog, or tiny red mites on plants. Are they insects?

To find the answer, let's look at a typical example of an insect—the butterfly. Think of the ways in which the butterfly is different from a spider. First there are the big wings. Of all the crawling creatures, only insects have

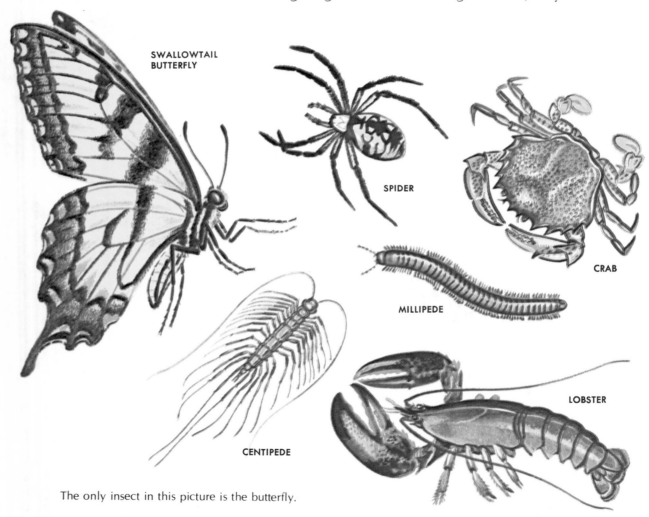

The only insect in this picture is the butterfly.

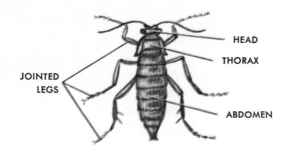

The wingless spanworm shows that the butterflies have the same distinct parts that all insects have.

wings. Although spiders may sometimes sail through the air at the end of a long, thin silk thread, no spider can really fly.

Count the number of legs on a butterfly. You will find that there are six legs. A spider has eight. Crabs and lobsters have ten. Other creatures may have even more. But an insect has only six legs as an adult. Some infant insects seem to have too many legs. Others, like fly maggots, seem to have none at all.

Another way to identify an insect is to count the number of main body parts. When you look at a butterfly, you can see that it has three main body sections:

(1) A head, with the antennae or feelers.
(2) A chest or thorax with all the wings and legs.
(3) A tail section or abdomen.

The spider seems to have only two parts. Crabs seem to have only one. Scorpions and centipedes have many. And they all have many legs and no wings. So they are not insects.

Not all insects have wings, either. Fleas, some crickets, and even some beetles and moths cannot fly at all. But they still have the right number of legs and body parts for insects—six legs and three main body sections. In fact, the name insect means "in sections."

Nobody knows exactly how many kinds of insects there are, but we are sure that there must be more than a million different kinds. Some scientists think there may be seven or eight million kinds—perhaps even more. But we do know that there are more kinds of insects crawling and swimming and flying around than all the other kinds of animals put together.

Each of these insects has its own interesting story. One kind of wasp makes jugs of mud that bake so hard in the sun that they look like stones. Some ants raise plants in tiny gardens. One kind of fly catches a mosquito and lays its eggs on it; then, when the mosquito bites a person, the fly maggots drop off and burrow under the person's skin.

There are insects that look like sticks. One of them, the giant walking stick, may be more than a foot long and wider than your finger. It is brown and scaly-looking, like a branch. Its six legs and two antennae look like twigs.

Some insects look like plant parts. Have you ever chased a bright orange butterfly in the woods? Its colors may be seen many yards away. Just when you think you have it, it disappears. No matter how hard you look, you can't find it. Then it suddenly flies up from right under your feet. If you catch it, then you know why it has been so hard to see. Its bright wings are the color of a dead leaf on the underside. When it folds its wings, the underside is all that shows. It looks like an old brown leaf.

Many moths can hide in plain sight on the trunk of a tree. Their speckled color is just like that of the bark. A long-legged water bug looks like a floating wisp of hay. Some green insects are shaped just like leaves, while others look like flowers.

One of the most interesting stories is that of the seventeen-year cicada. It lives seventeen years in darkness below the earth as a young, undeveloped insect called a nymph. Then suddenly, millions of them come out at once. They leave little holes in the ground about the size of a dime. They cluster so heavily on bushes and trees that the branches bend down with the weight of what looks like large dark berries.

Sometimes insects use tools to help them with their work. One wasp picks up a pebble and uses it to pack the earth on top of its eggs. A certain ant uses its babies just as you would use a tube of glue. The ant picks up the baby

This dead-leaf butterfly looks like the leaves near it.

Insects always carry their tools with them.

CRICKET DIGGING

ICHNEUMON DRILLING

PRAYING MANTIS CUTTING

and presses it against the edges of a curled leaf. The sticky material from the baby's mouth glues the leaf edges together. The ant lion sometimes throws pebbles into the air so that an insect may be knocked down into its pit—and eaten.

Did you ever forget where you left a hammer or shovel? This couldn't happen to insects that have their tools with them at all times. The mole cricket has large feet that look like shovels. Burrowing beetles have "shovels" on the end of their snouts. They are just right for digging the soil. Water striders have "waterproof boots" in the form of large legs and feet that enable them to run around on top of the water without getting wet. Diving beetles have little air pockets so that they can breathe under water like skindivers.

The praying mantis has spiny legs that open and close like a jackknife and hold its food tightly. A fly can walk upside down on the ceiling because of special pads and hooks that hold it in place. The ichneumon fly has a long drill at the end of its body, with which it can drill deep into a tree trunk to lay its eggs in the hole of a wood borer. The tiger beetle has stiff hairs on its feet so that it can run over sand without slipping.

Seventeen-year cicada nymphs. Cicadas include 75 different species.

Bees carry many tools. They have "combs" and "brushes" on their legs, which help them work with the wax of the hive. They have a "basket" to carry the pollen from flowers. Wing hooks keep their front and hind wings together when they fly. These become unhooked when the bee folds its wings.

You can find insects nearly everywhere you look. Mountain climbers find them on high peaks. Explorers bring up blind white crickets from deep caves. Little gray insects called springtails skip about on winter snow. Their dark-colored bodies soak up the warm sunshine and keep them from freezing.

One kind of insect lives right on the edge of Niagara Falls. It is kept from being swept over the falls by a strong thread holding it in place. Other kinds live only in the still water of ponds. Many live inside the stems of weeds. Some fly high into the air, while others spend their lives within a few inches of where they were hatched. If you look at the skin of an orange, you may see some tiny brown scales. These are scale insects, and they don't move at all. Other scales move very little.

Some insects live under rugs and furniture. They may sometimes find their way into your breakfast cereal. Termites and carpenter ants may tunnel through the boards of your house. One little creature seems to like books; it spends all of its life in libraries.

There is one great place on earth where insects are not found, and that is the ocean. Insects have never been able to do very well in the seas. Their bodies cannot get used to the salt water. Only a few kinds go into the sea at all, and these stay right near shore. So, even though there are millions of insects, they are crowded close together and fenced in by the oceans that surround us.

Life in a Suit of Armor

If you cut open an insect, you wouldn't find any bones, no matter how hard you look. Its skin is the only skeleton an insect has. Without it, the insect would be soft and helpless.

Flies and mosquitoes have thin skins. The beetle looks like a knight in armor with its thick heavy shell. Even soft aphids live in a thin jacket.

If you wore a space suit that covered your hands and face, how would you be able to feel and smell? You would need little holes to sniff through and other holes for your fingers to feel through. Insects have tiny hairs that poke out through the armor. They also have little pits and pockets. These hairs and pockets help them smell and touch and taste.

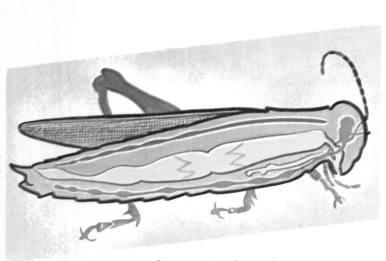

Cutaway view of a grasshopper.

Bird taking a dust bath.

Sometimes these pockets and hairs are on the legs of the insect. Many of them are on the feelers or antennae. They may be on other parts of the body. So we can say that some insects smell with many parts of their bodies. In fact, insects don't have noses at all.

Use a magnifying glass to look carefully along the sides of a large insect. You will see a row of circles that look like the portholes of a ship. These are the breathing pores. They are called spiracles. Instead of breathing through noses, as we do, insects breathe through these holes.

The spiracles lead to little tubes. These branch all over the inside of the body, even in the legs and eyes. When the insect moves, air is pumped in and out. Even water insects have these tubes. They help the insects get their oxygen from the water around them.

Birds take dust baths to suffocate insect pests in their feathers. The dust clogs up the insects's spiracles, and since they cannot breathe, they die.

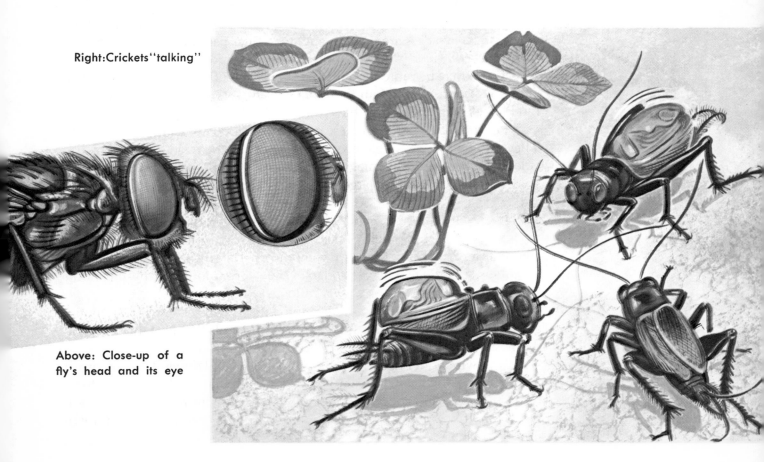

Right: Crickets "talking"

Above: Close-up of a fly's head and its eye

When we talk or sing, the sound comes from our throats. A "singing" insect makes its noise by buzzing or scraping. Crickets rub their wings together. Grasshoppers rub their legs and wings together. Cicadas have a "drum" on their bodies. Other insects scratch their bodies or grind their jaws to make a noise. They find each other by following these noises. Sometimes they use the noises to frighten away their enemies.

When insects fly, their wings make a humming sound. Sometimes the muscles of the insect make a hum, as well. The higher the hum, the faster the wings are beating. A buzzing housefly beats its wings twenty thousand times a minute.

Katydids have little patches on their legs that are sensitive to noise. Grasshoppers have their ears on their abdomen. Some insects can feel sounds or vibrations through their feet, just as you can feel a radio playing by touching it with your fingers. Scientists have not yet found the ears of the champion noisemaker of them all, the cicada. As far as they have been able to discover, it has no ears. It seems to make all that noise for nothing.

If you think insects have just two eyes, you have guessed wrong. What appear to be two eyes are really many small ones packed together. They are called facets.

There may be more than fifty facets in each of the two large eyes of an ant. One scientist found four thousand in each large eye of a housefly. Some moths and dragonflies may have a total of fifty thousand facets.

For many insects, even these great numbers do not seem to be enough. They also have a few single eyes right in front of the head. These look like colored beads. Like little magnifying glasses, they probably help the insect see things up close. The large compound eyes help the insect see things farther away.

Even with all those eyes, insects cannot see very well. They depend mainly on taste and smell. Those little hairs and pits on the body and antennae are very keen. Many male moths have large feathery antennae that help them find the female in the dark. One scientist found that some moths can find another moth as far as a mile away.

If you have a chance, watch some ants at work. They feel along the ground with their antennae, following definite trails that lead them back to the nest. Each ant follows the trail of the one ahead of it.

Now wipe your finger hard across the trail several times. This will brush away much of the scent. Then watch the next ant that comes along. It stops, turns in circles, and goes from side to side. It seems completely lost, even if the nest is only a few inches away. It may take three or four minutes for it to find its way again.

"Follow-the-leader." Ants following the trail of the one in front going to the nest.

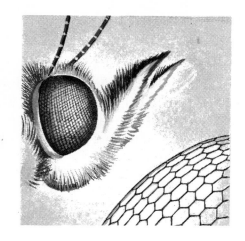

MALE BAGWORM MOTH

The compound eyes of a moth are made up of hundreds of six-sided lenses (far left). Left: the feathery antennae of a Cynthia moth.

One scientist saw a long line of caterpillars. Each was following the one ahead of it. The line went over logs and under bushes, like a little train.

Then he had an idea. He put some of the caterpillars on the edge of a glass bowl. Around and around they went, following each other's trail in a circle for days and days. They never stopped or climbed down. They just kept on playing follow-the-leader until the scientist took them off.

This was more than just a game for the scientist. He was finding out some important things about insects. Other scientists were also studying them.

Beekeepers know that bees find their way by means of the sun. But what do they do on a cloudy day? They can still sense where the sun is in the sky by means of polarized light. This is light that can be seen better from one direction than from others. Even behind the clouds, the sun still sends it down. Ultraviolet light (the same invisible rays that cause sunburn) also guides the bees. And their huge compound eyes make out the shape of familiar trees and houses. They find their way by the colors of flowers, too—except for one color. Bees are color blind to red.

Scientists have found that insects know how to act as soon as they are born. We have to learn to nail boards together, but insects can make perfect homes on the first try. Our parents help us decide what food to eat, but insects usually never see their parents. The hungry babies know what to eat as soon as they hatch. They know how to hide their eggs and to keep out of danger.

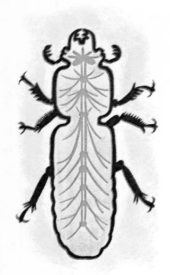

An insect's nervous system is made up of nerve centers called ganglia. They run lengthwise, and a double chain of nerves connects them. Nerves branch from each ganglion to other parts of the body. The large pair of ganglia in the front is called the brain.

They can do these things because of what we call instinct. This usually helps the insect to meet all its problems. Instinct tells a Japanese beetle to drop to the ground out of sight the minute you touch its twig. Instinct tells a bombardier beetle to wave its abdomen in the air and squirt you with an ill-smelling spray, like a little skunk. Instinct helps a squash bug put its eggs where they will be hidden and yet near the best food.

Instinct is some kind of inner knowledge that helps the insect to do something although the insect has never been shown how to do it. So instinct may be described as built-in or ready-made knowledge. Scientists, however, don't know what causes it or how it was made.

Insects Helpful to Man

Although many insects eat our gardens and forests, some kinds are useful to us. Perhaps you have watched a pair of burying beetles as they dug under a dead mouse until it sank into the ground out of sight. Maybe you have seen ants cleaning up some garbage by taking it into their nest. One kind even carries away cigarette butts.

Ladybird beetles eat plant lice. Some kinds of stinkbugs feed on harmful caterpillars. Water striders keep the water clean by feeding on insects that drop from bushes. Hornets fly around cows and horses, chasing flies until they catch one for food.

A Pacific island native with a live butterfly in her hair.

With its long proboscis, the bee sucks the nectar that is produced in the bottom of the flower. In doing so, the bee pollinates the flower, making it possible to produce seeds (above).

The bees collect pollen on their hind legs. Pollen is held in place by long leg hairs, which form the "pollen baskets" (far right).

BEE (ACTUAL SIZE)

EMPTY POLLEN BASKET

FILLED POLLEN BASKET

Perhaps you have seen a painter using shellac. It looks like varnish and is used on boats and airplanes. It comes from the lac insect of India. Some bright colored dyes are also made from insects. Silk is made by silkworms to cover their cocoons. The Chinese keep singing crickets in little cages. The ground-up bodies of some insects are even made into medicine.

One of the strangest uses for insects is that of lighting a room. In many parts of the world there are no electric lights. When the natives in some tropical countries want to see after dark, they go outdoors with a little cage. They put a few fireflies into the cage. Each firefly has a spot in its body that glows when air is let in through the insect's air tubes. The shining of a dozen large fireflies helps light the room. Some native girls even wear a butterfly in their hair.

The more circles a bee "dances," the farther away are flowers with nectar (above).

The straight line in the bee's dance shows the direction of the flowers. The tail wagging gives the scent (above right).

Carefully taste a drop of liquid found in the bottom of a flower. It is used by bees in making honey. When the beekeeper takes the honeycomb from the hive, he always leaves plenty for the bees to eat during the winter. Otherwise they would starve.

When bees go from one flower to another for the sweet nectar they make into food, they also pick up some pollen on the hairs of their body and legs. A little of this pollen brushes off as they visit each new flower. It helps the flowers' seeds and fruit to grow. Without bees, plants couldn't produce apples, peaches, melons, and other good things we have to eat.

Bees have a special way of communicating with each other when one of them has found a source of nectar. This was discovered by the German scientist Dr. Karl Erisch when he solved the mystery of the so-called dance of the bees. When a worker bee returned from an apple tree, she began to do a little dance near the entrance to the hive. First she circled one way, then another. In between the circling she walked a little straight line, wiggling like an excited puppy. Soon the other bees followed her in her dance, doing just what she was doing. The circles tell how far away the flowers are—the more circles, the farther away. The straight line tells the direction to travel, and the odor of the flowers still clinging to her body tells them what kind of flowers they will find. In a few minutes they fly away, one after another—right to the apple tree!

Eggs by the Thousands

Insects have many enemies, and it is not surprising that they have found many ways to protect themselves. Each kind has its special way of caring for itself.

Even the eggs are given special care by the mother insect. How often do you see any insect eggs? If you were able to count all the eggs within a few miles of your house this minute, you would find that there were millions of them. Yet you might look for a long time before you could find any at all.

Some insect mothers bury their eggs deep in the soil. Grasshoppers poke the end of their bodies down as far as they can reach and lay the eggs in a hole. Some beetles dig down out of sight to lay their eggs. Ants and termites

Grasshoppers poke their abdomens into the earth as far as they can reach to lay from 20 to 100 eggs.

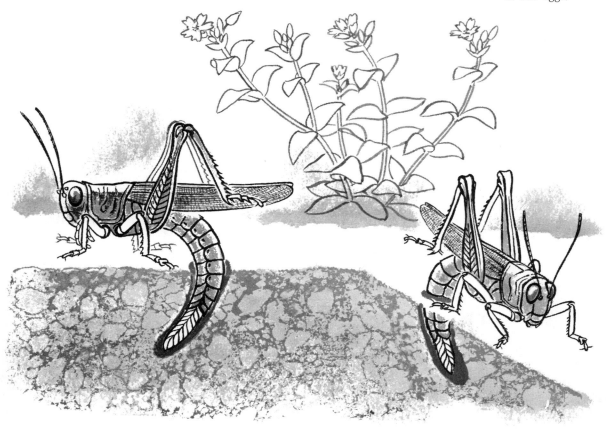

The Viceroy (right) looks like the Monarch (left).

CECROPIA MOTH

have nests under a stump or in a mound of earth. There the eggs are safely hidden and protected from enemies. Some insects produce a liquid into which they put their eggs. Later, the liquid hardens and the eggs are safe in a covering.

Sometimes you can find insect eggs on leaves and twigs. They may have tough shells so that other insects cannot eat them. They may be covered with wax to protect them from winter winds. Perhaps you have seen the egg case of a praying mantis. This fluffy case is like a blanket in the snow. The eggs are safe inside.

You may find a green twig that looks as if someone had been cutting it with a knife. Possibly you will find an insect egg at the bottom of each cut. A cicada makes the cuts with the sharp tip of its body. Then the eggs are safely hidden under the bark.

There are many other places where you can find the eggs of insects. Flies lay their eggs in garbage. Lice attach their eggs to the hair of animals with a special glue of their own. Some walking stick insects drop thousands

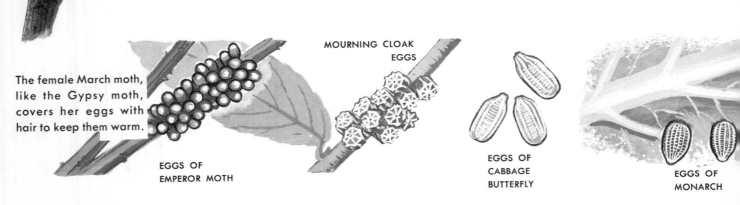

The female March moth, like the Gypsy moth, covers her eggs with hair to keep them warm.

EGGS OF EMPEROR MOTH

MOURNING CLOAK EGGS

EGGS OF CABBAGE BUTTERFLY

EGGS OF MONARCH

of eggs from the trees; it sounds like falling rain. Clothes moths lay tiny eggs in the wrinkles of coats and suits and other woolen garments.

Mosquitoes lay large numbers of eggs on the water. Examine the underside of a water lily leaf. You'll find many kinds of eggs. Perhaps you have seen a dragonfly darting along over a pond. It dips down every few seconds to drop an egg beneath the water.

Some damselflies fasten themselves together. Then the female goes beneath the surface of the water to lay her eggs while the male stays above. When the eggs have been laid, the male pulls the female out of the water.

One female water bug makes the male bug take care of the eggs. She catches him and lays her eggs on his back!

Some of the eggs are round. Others are flat. Some are brightly colored. Others are wrinkled and brown. Many are black. There are eggs shaped like the crown of a king. Others look like little jugs with pop-up lids. If you have a magnifying glass, you can see various shapes and sizes of insect eggs.

Some tent caterpillar eggs take two years to hatch. Fly eggs may hatch in a few hours. Many eggs laid in the fall will not hatch until spring. Some eggs hatch inside the mother insect, so that tiny insect babies are born.

Day after day the queen honeybee lays thousands of eggs—often four or five thousand a day. Such egg production is a great task, so she is fed by the workers almost constantly. They feed her and lick her, caring for her every need. The workers also take care of the eggs, but queens of many other kinds of bees take care of their own eggs. One of the strangest is the queen bumblebee. She sits on her eggs in the spring like a little bird. All winter long she remains in a sheltered spot, hidden from the cold and storm. When spring comes at last, she goes house-hunting. She flies over the meadow, looking for a little hollow in the ground. Sometimes she even uses an old mouse nest or a half-hidden tin can.

She makes two wax cells. One of them is her honeypot. She fills it with

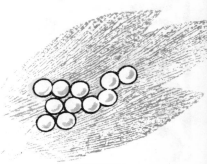

Insects lay eggs of various shapes, sizes, and colors. All the insect eggs pictured here have been enlarged.

SWALLOWTAIL BUTTERFLY

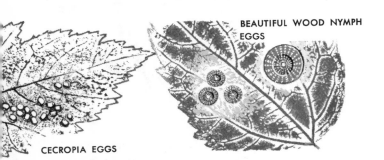

CECROPIA EGGS

BEAUTIFUL WOOD NYMPH EGGS

HARLEQUIN BUG

Oak gall wasp on leaf.

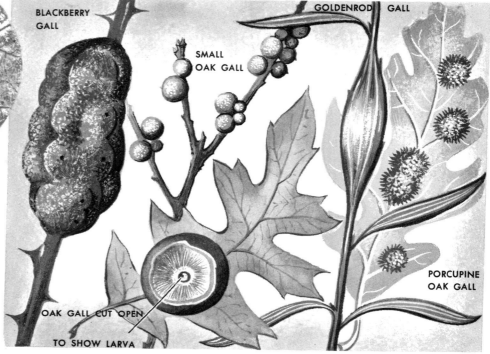

Some typical insect galls—swollen plant tissue—caused by the hatching of wasp eggs.

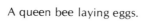

A queen bee laying eggs.

The gray, comma-shaped egg is attached to the bottom of the cell with an adhesive secretion.

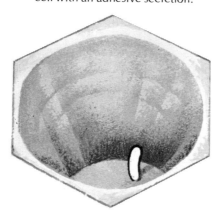

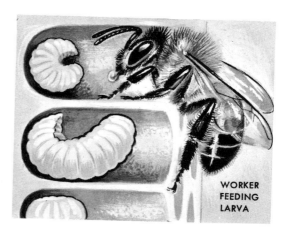

A blind and legless grub hatches from the egg. It is fed continuously until it starts spinning a cocoon after a few days.

honey she makes from the early spring flowers. The other is her brood cell. About eight eggs go into it, often on a little pad of pollen that will serve as food. Then she caps it over with wax and settles down on top of it.

For the queen bumblebee, babysitting is no problem. When the insect babies (called larvae) hatch in three or four days, she opens their cell to feed them. But when they are fed, she closes it up again. She may do this several times a day, flying away for more food in between.

The female gall wasp has another way of providing a safe place for her eggs. She pokes her eggs into a plant leaf or stem, which then forms a gall or swelling. The gall serves as a home for the wasp larvae, and the larvae feed on the material inside it. Often they spend the entire winter in this shelter.

Moths and butterflies go through many changes in their short lives. The eggs laid by the females hatch into little caterpillars. The caterpillars

Here is a "damsel train" in action. The female damselfly lays her eggs in the water or on water plants.

23

later turn into pupae, which will finally become adult moths or butterflies. This whole process of change is called metamorphosis.

Each moth and butterfly has its own special type of egg. The gypsy moth lays a sticky mass of round eggs and covers them over with the hairs from her own body. They stay under this blanket all winter. The tent caterpillar moths cover their eggs with a foamy mass of glue. This hardens into a tough covering that keeps out snow and water.

Some spiny-tailed moths poke a hole in a twig and lay their eggs inside. Many moths and butterflies just dart at leaves and fly away, but they leave their moist eggs clinging to the plant.

The number of eggs laid differs with different types of butterflies. Usually, each butterfly lays two or three hundred eggs, but only a few hatch. Many eggs laid in the fall do not hatch in the cool fall weather. They wait until spring. Sometimes, however, the weather stays warm for weeks. If the caterpillars began to crawl around on a warm November day, they would soon starve for lack of food. So nature has provided a wonderful safeguard to keep them from hatching too soon.

The safeguard works in this way. Unless many eggs first go through a period of freezing, they will not hatch, no matter how warm the weather. Thus, a cold winter day actually helps them get ready for hatching in the spring.

Not all eggs take months before they hatch. A few days after the cabbage butterfly has laid her single white eggs on the leaves of cabbages and turnips, the newly hatched green caterpillars are nibbling at their salad. The eggs of certain other moths take even less time. In fact, they hatch while they are still in the body of the mother. When she lays her "eggs," they have already become tiny living caterpillars.

Many caterpillars can eat only certain plants. The mother seems to know which plants will make good food. She probably uses the sense of smell in her delicate antennae. She lays her eggs on plants that give off the proper odor. If she cannot find the exact plant, she will pick a close relative. Thus a clover moth may also lay her eggs on alfalfa, and a bean moth may lay her eggs on peas.

Scientists have found that they can fool moths by tricking their sense of smell. Cabbages, turnips, radishes, and broccoli all belong to the mustard family. By smearing mustard oil on paper, they can make cabbage butterflies lay their eggs on it. The corn earworm moth may lay her eggs on cloth soaked in corn oil, instead of on corn silk.

The tobacco sphinx moth usually lays her eggs on tobacco plants. But if

Eggs of Violet Tip butterfly.

Eggs of American Tent Caterpillar Moth.

The eggs of the bagworm are in the "cradle" made by the larva of this moth.

24

Adult butterflies and their larvae.

a pack of cigarettes is available, she may carefully place her eggs on the outside of the package.

Some of the eggs rest in strange cradles. Peacock moths lay their eggs on stinging nettles. Cactus borer moths crawl over prickly spines to lay their eggs. The wax moth creeps into a beehive, even though the bees would kill her if they found her. Her newly born caterpillars will feed on the beeswax in the honeycomb.

The female yucca moth pierces the flower of the yucca or Spanish bayonet. She then lays a few eggs with her sharp egg-laying organ and rolls a ball of pollen from another yucca flower around it, which will help the flower to form seeds. The seeds grow just in time to feed the new caterpillars.

Scientists know of no other way in which pollen can get from one yucca

A cicada killer buries its victim in an underground passage. The paralyzed cicada has been stung and will provide food for the larva of the cicada killer when the egg hatches later on.

flower to another. Without the little white yucca moth, there would be no seeds. Since the caterpillars do not eat them all, there are always some seeds left over to grow new plants.

Bagworm eggs are in one of the strangest cradles of all. The female caterpillar makes a silken pouch and covers it with evergreen needles or twigs. Wherever she goes she pulls this pouch around, like a turtle with a shell. When she turns into a moth, she has no wings or legs and stays right in her pouch. Soon after laying her eggs, the female dies. When the new caterpillars hatch from the eggs, they must make their way out through their mother's pouch.

Insect Babies—All Shapes and Sizes

Moths and butterflies lay eggs that hatch into caterpillars. Large, buzzing bumblebees have little grubs for babies. So do beetles and wasps. Fly eggs hatch into maggots. Caterpillars, grubs, and maggots are called larvae.

Grasshoppers and dragonflies have babies that look a lot like the parents. They have little buds where their wings will grow some day. Their heads seem too large for their bodies. These insect babies are called nymphs.

There is one way in which all these different babies are alike. They are nearly always hungry. They begin to eat soon after they hatch and keep on eating for most of their lives. So the eggs are laid where the insects will have food as soon as they hatch.

Perhaps you have seen a wasp pulling and tugging at a caterpillar that had been stung so that it couldn't move. The wasp will poke it down into a new hole in the soil where she has laid her eggs. The new wasp babies will then have food to eat when they hatch.

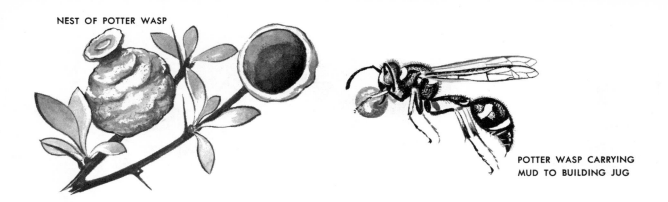

NEST OF POTTER WASP

POTTER WASP CARRYING MUD TO BUILDING JUG

Some new insect babies are so hungry that they will eat anything at all—even their own brothers and sisters. But the lacewing fly has solved this problem. She lays each egg at the end of a long stalk. When the fierce little baby hatches, it drops off the stalk and begins to hunt for food. Its brothers and sisters are safe on their stalks above.

New baby insects can protect themselves, even in a world filled with hungry enemies. Many of them are the same color as the leaves they eat, so that they are hard to see. Some have fierce-looking spots that make them seem to have great, round, dangerous eyes. Some have sharp spines, making them look like tiny cactus plants.

One insect puts out a pair of ill-smelling horns when it is in danger. Some insects are long and brown and look just like a twig. Others are round

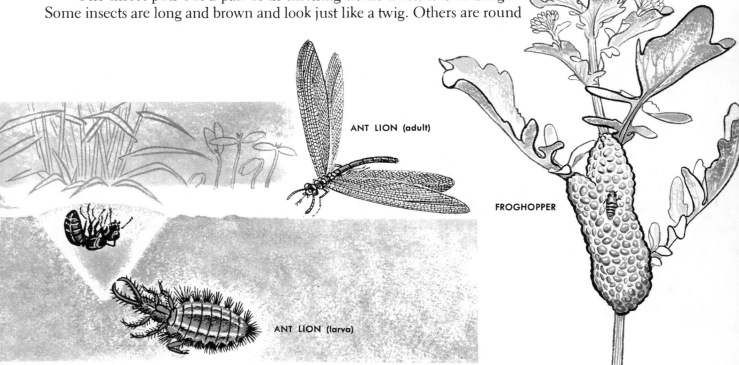

ANT LION (adult)

ANT LION (larva)

FROGHOPPER

27

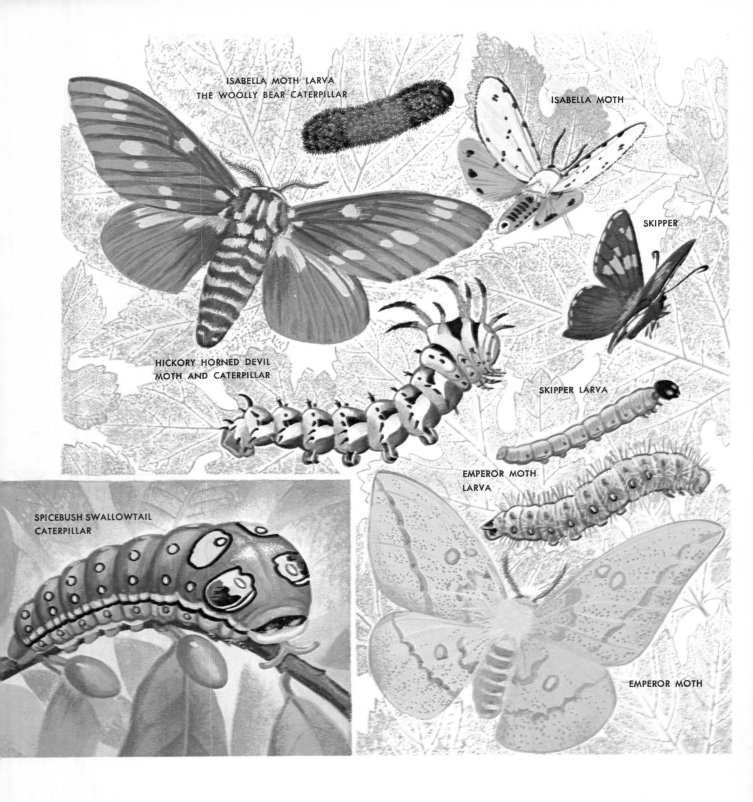

The larva of the codling moth (left) is a widespread apple pest. Above, scorpion fly sucking on a larva.

and gray like a pebble. Sharp-jawed ones may pinch you if you bother them. Others curl up and drop into the grass at the slightest touch. Still others are poisonous. Their enemies soon learn not to eat them. Baby insects find protection in their shapes, colors, odors, body poisons, and fierce looks.

Most insect babies have no parents to take care of them. The adult insects, their mothers and fathers, usually die soon after the eggs are laid. Wasps, bees, ants, and termites, however, take good care of their young. They build nests with many caves and tunnels. Here they have rooms that may be compared to our nurseries, kitchens, and storehouses. These nests may be many feet high, and some are twice as tall as a man.

Many baby insects build homes of their own. The caddisfly larva lives on the bottom of streams and ponds. It makes a tube of sticks or sand grains glued together. Then it fits itself inside the tube. It looks like a little turtle as it bumps along the bottom of a stream.

One of the strangest homes is the bubble house of the froghopper. You can see many of these nests on grass blades and weeds. If you poke inside the bubbles, you will find a little green froghopper. Put it on a new blade of grass and it will begin to blow bubbles until it is hidden.

The ant lion makes a pit in the dry sand. It waits at the bottom of the pit with its pincers open wide. If an ant stumbles into the pit, the ant lion has its dinner.

Some caterpillars make webs to protect themselves. Other insect babies roll up leaves or cover themselves with dust. Some tiny insects even tunnel in the leaf of a tree, leaving strange marks. Once people thought that the trails of the leaf miners were the writing of ghosts.

The baby insects keep on eating and growing. But they don't grow as we do. An insect's skin doesn't stretch to make more room. It becomes tighter and tighter, like last year's jacket.

One day it splits along the back, and the young insect crawls out of its old skin. Its new skin is soft and thin, and its body swells up quickly. Soon the new skin hardens. Then the insect can no longer grow until after it splits its jacket again.

Only young insects can grow in this way. When a caterpillar turns into a moth, or the grub becomes a beetle, they will never shed their skin again. They stay the same size for the rest of their lives. Little moths don't become big moths, nor little flies big ones.

Some insects make so much noise when they eat that you can hear them. Perhaps you have heard a scratching sound coming from a wood pile in the forest. It may have been a family of wood borers, a kind of beetle grub. You can often hear them chewing away.

Maybe you have read, in the Bible, about locusts that attacked crops in ancient times, or in the newspapers, about locust attacks in more recent years. Millions of locusts eating a field of grain can be heard some distance away. They sound like the wind in dry leaves.

A cricket or beetle grub chews its food. But some insects sip their food quietly through a long tube. They drink the sap of plants or the blood of animals. If you look where their mouths should be, all you see is a long, pointed tube. Think how it must be to go around with your mouth shut tight and just a straw sticking out!

Who eats more food—you or your parents? Many growing insects eat much more than their mother and father eat together. They may eat more than their own weight in food each day. They are growing so fast that they never seem to get enough food.

The Unlucky Caterpillar

Have you ever watched a caterpillar on the sidewalk? Maybe you've seen it chewing on some leaves. If you have a garden, perhaps you have helped spray the plants so it wouldn't eat them.

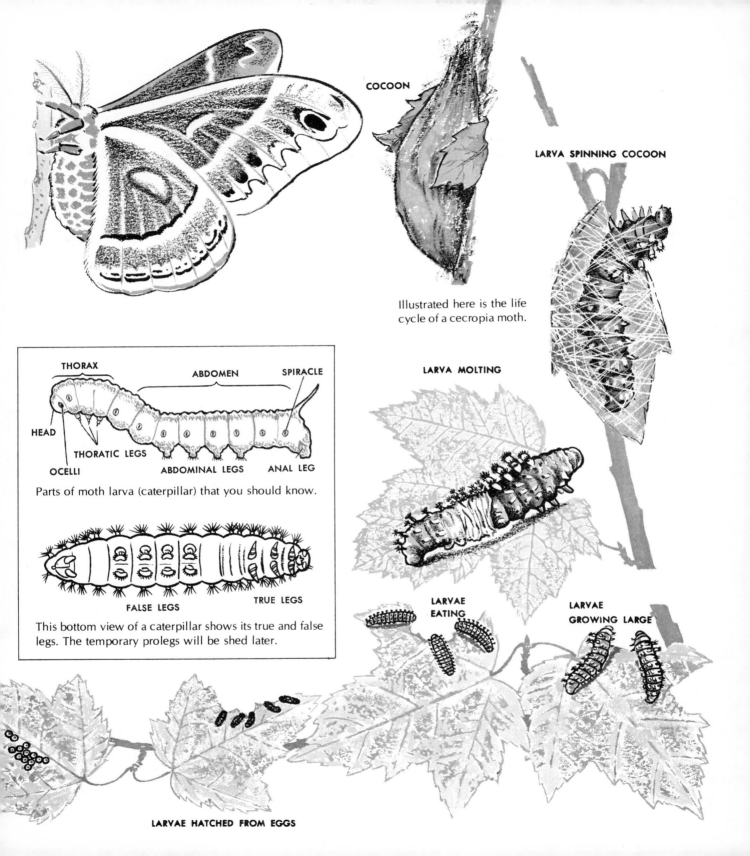

When the caterpillar eats leaves that have been sprayed with a poisonous chemical, the poison may kill it. At the very least, the poisonous chemical will cause it to move away.

The caterpillar has been in danger ever since its butterfly mother first laid her eggs. No matter how carefully the eggs are hidden, other insects come along looking for tiny bits of food and often find them. Storms and cold weather kill many caterpillars in the eggs, too.

When the egg hatches, insects and spiders are waiting for the caterpillar's appearance. Birds look at every leaf and twig, eating every caterpillar they can find. Snakes, lizards, toads, and frogs catch more of them. And when you go out into the field to make an insect collection, you will catch some, too.

Look at the head of the next caterpillar you find. It seems to have two great round eyes in front, but they are not eyes at all. The real eyes are little pinpoint dots that can hardly be seen. It can see only a few inches ahead. Probably the only way the caterpillar knows danger is near is when the leaf shakes as a bird lands near it, or when it smells the scent of a nearby enemy.

A caterpillar's body is divided into thirteen ringlike parts called segments. Attached to the three segments nearest the head, in the thorax body section, the caterpillar has six small, stubby legs. It also has a few extra pairs of legs along the sides of the segments that make up the abdomen body section. These false legs, called prolegs, are temporary, and the caterpillar will shed them along with its last skin.

You may like a snack after school, or just before going to bed, but the caterpillar is always hungry and eats almost all day and all night. It needs a great deal of food because it is growing so fast. So it keeps munching on the turnips and radishes in the garden. If the caterpillar has to go without food for more than a few hours it will starve.

As a caterpillar eats, it continues to grow. However, like other insects, the caterpillar's skin doesn't grow as the rest of its body does. It remains the same size, and finally the tight skin splits. The caterpillar then sheds its skin by wriggling out, a process that is called molting. But underneath is another skin. The caterpillar will outgrow this one, too. In fact, it will molt several times before it reaches the end of the caterpillar stage and is fully grown.

There is a soft "plop, plop" in the forest. It sounds as if the trees are dripping after a rain. It is really the sound of larvae dropping to the ground. In other trees, larvae are lowering themselves slowly on their silken lifeline. Some larvae crawl down the stems of plants or burrow into the ground.

They are no longer interested in food. The time has come for them to enter their next stage of life and become a *pupa*. Some caterpillars may take days to find just the right spot. Other kinds merely settle down in the fork of a tree. Not all caterpillars enter the pupal stage at the same time. New pupae develop every day as new larvae become old enough.

Insects develop in one of three different ways. At right is the life cycle of a silverfish. This insect looks like its parents as soon as it hatches from the egg, and it grows gradually until it reaches adult size. This is called gradual development. Below left is the life cycle of the dragonfly, which has an incomplete metamorphosis or development. Metamorphosis means "change of form." The butterfly shown below right has a complete metamorphosis. Its life cycle includes four stages—the egg, larva, pupa, and adult stages.

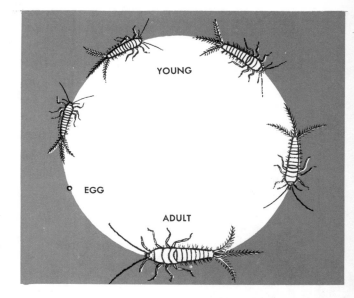

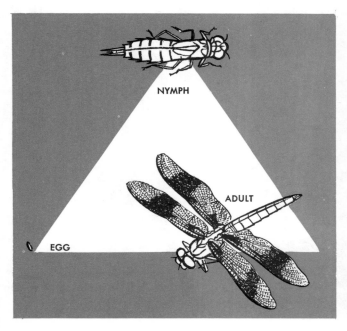

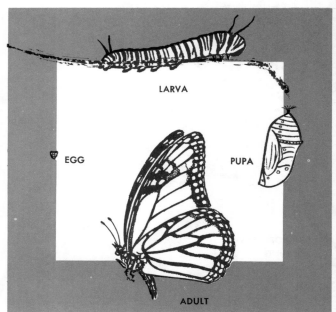

The "Sleeping Puppet"

Many moth caterpillars make a cocoon, whereas butterfly caterpillars make a similar structure, called a chrysalis. To build its cocoon, the caterpillar often starts by making a soft, silken "mattress." Standing on this, it weaves its head back and forth, around and around. With each movement, the strand of silk is pulled from the silk gland on its lower lip. Gradually, the wall becomes thicker. Finally the busy little spinner can hardly be seen inside its little room.

This is one way in which the coccon is made. There are about as many other ways as there are moths. Many cocoons are beautifully formed, with ribs and cross-strands like fine lace. The cecropia moth caterpillar makes a cocoon of silk so tough that it is almost impossible to tear it. It may be larger than a hen's egg.

The fall webworm and the Asian silkworm make silk suitable for thread; the silk of other larvae is less useful to us.

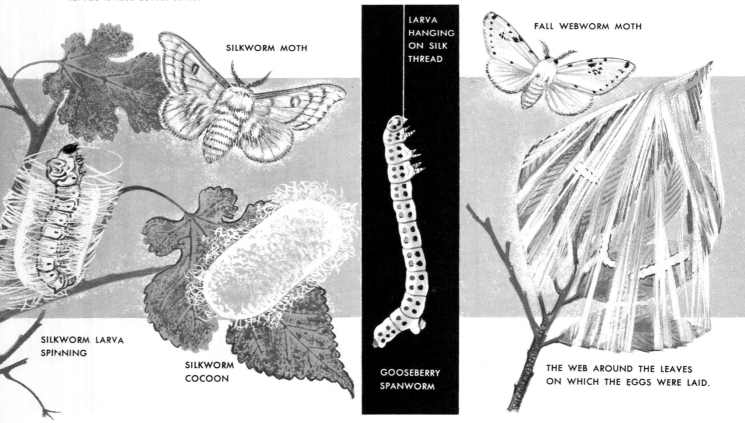

SILKWORM MOTH

SILKWORM LARVA SPINNING

SILKWORM COCOON

LARVA HANGING ON SILK THREAD

GOOSEBERRY SPANWORM

FALL WEBWORM MOTH

THE WEB AROUND THE LEAVES ON WHICH THE EGGS WERE LAID.

Other moth caterpillars roll themselves in a leaf. Bits of dirt, bark, and even the old larval skin may be woven into the cocoon. These materials help to hide and protect them. Once the caterpillar has become a pupa inside its cocoon, it is completely helpless.

The pupa of a butterfly is not covered with silk. Usually it is very hard and shell-like. It may be shaped like a twig or a piece of bark so that it is hard to see. The name chrysalis means "golden." Often butterfly chrysalises are gold in color or have gold beads and ornaments.

Just before its last molt, a change comes over the caterpillar. It humps its back and pulls in its legs. If it becomes a chrysalis, it may hang down from a twig, holding on by hooks at the end of its body. If these lose their grip, it falls helplessly to the ground.

Finally the larval skin splits for the last time. Out pushes a strange new creature. Blind, with no legs for walking or wings for flying, it can move only by twisting its abdomen. It is a pupa at last, protected by the silken cocoon or the shell-like covering.

The name pupa means "puppet" in scientific language. If you look

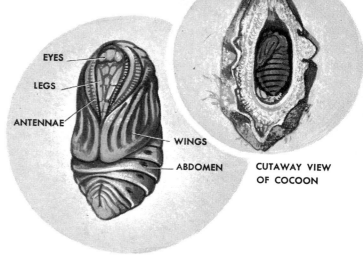

The parasitic wasp egg has developed inside the pupa of a butterfly and now hatches (left).

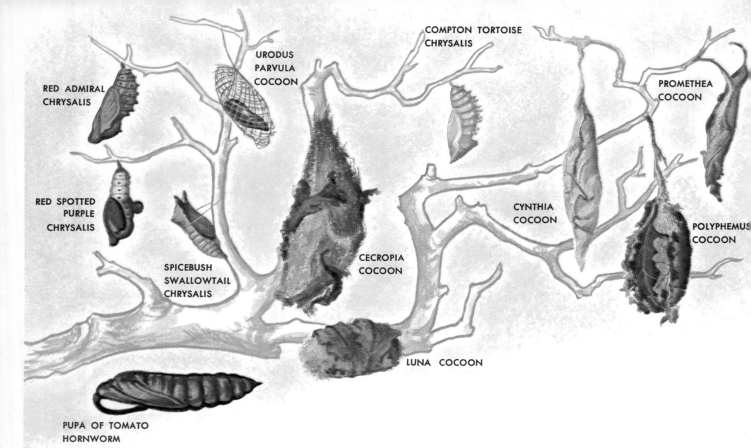

closely at a pupa, you will see that it looks like a little doll or puppet. It may be compared to a mummy, because the outlines of the new legs can be seen as if they were tightly wrapped in bandages. You can also see the wing pads and the covers of the long mouth and antennae.

Now the pupa can do nothing but wait. It cannot take any food. It cannot drink. If even the smallest ant nibbles at it, it has no defense except its heavy coating. If dust gets into its spiracles, it may suffocate. It is as helpless as a person tied with rope from head to foot.

Some sphinx caterpillars burrow beneath the ground. There the dark pupa, looking like a little jug with a handle, is hidden from sight. The puss moth pupa is covered over with the stinging hairs of the larva, so it is safe. Many pupae are tucked beneath bark, under stones, and under the edges of roofs. Some hang right out on a branch all winter, wrapped in an old leaf. They look so natural that nothing bothers them.

The little wasps that attack caterpillars sometimes find the pupa also and lay their eggs in it. Then, instead of a new moth emerging from the cocoon, several dozen wasps hatch.

One African moth, however, fools its enemies. It makes a cocoon with false bumps on the outside so that its moth cocoon actually looks like a wasp cocoon. A wasp, seeing these bumps, flies away. Perhaps she feels that another wasp has already laid her eggs there.

Although a pupa seems quiet, it is not really sleeping. Inside, a

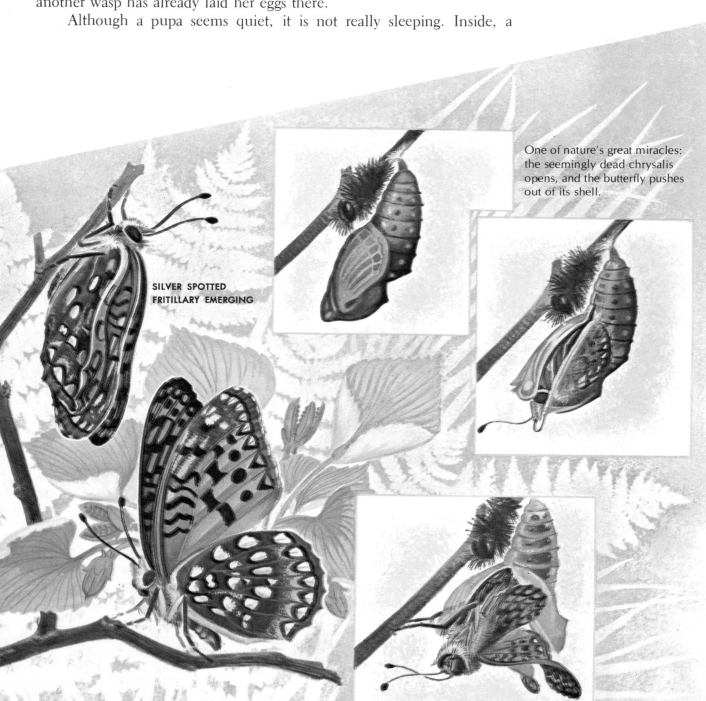

SILVER SPOTTED FRITILLARY EMERGING

One of nature's great miracles: the seemingly dead chrysalis opens, and the butterfly pushes out of its shell.

wonderful change is taking place. Entomologists, scientists who study insects, still do not know all that happens. Many of the muscles and organs used by the larva will not be needed in the adult. On the other hand, new ones will be needed. The old organs break down and supply their energy for new growth. Fat that was stored in the caterpillar body is also used for energy.

Some of the tissues, however, cannot be used at all. Yet if they stay in the body, they will be nothing but waste material. How does the pupa get rid of them?

The answer is found in tiny cells of the blood, called phagocytes. These act like little scavengers. They eat up all unwanted material and leave behind just what the insect needs.

PHOEBE FRITILLARY CHRYSALIS

MONARCH CHRYSALIS

Beetles and some other insects also form a pupa. You can tell the pupa of a moth or butterfly, however, by looking for the outline of the long, straight tube-mouth, or probiscis. Of course, if it is wrapped in silk, a pupa is quite likely to be that of a moth. Ant and wasp pupae may also be wrapped in silk, but they are usually found in the nests of these insects.

If you find a pupa and want to keep it for hatching, place it in a gallon jar with a wide mouth. A little paper toweling in the bottom will help to hold moisture. Supply several drops of water every few days, allowing some of it to splash on the pupa. However, do not allow the bottom to get too wet. Provide a branch or stick for the insect to climb on when it emerges from the pupa.

If the pupa is attached to a support such as a twig, bring this in, too. Then, when the moth emerges, it can crawl away and leave the pupal case behind. Otherwise, the case might stick to it when it moves around.

The process of change from pupa to adult may take months or even years. In a very dry year, some adults may not emerge at all, but wait for the following season. Others, such as the cabbage butterfly, may take only a few days. They may go through a whole generation in a month. Many pupae last from a few days to a few weeks.

No matter how long it takes, the time of pupation is finally over. Some cocoons have a little plug that the moth pushes out as it emerges. Others have a weak spot in the cocoon. Occasionally, the new insect has a silk-cutting ridge on its back or sharp edges on its wings. The pupae of one species have scissorlike jaws that cut through the cocoon.

Some pupae release a substance that softens and dissolves the silk in time for the delicate adult to leave without injury. Silkworms have this substance, so the pupa must be killed with hot water. Otherwise the fine strand of silk would be destroyed as the moth broke out of the cocoon.

When tortoiseshell butterflies emerge from their pupae, they release a red substance. Sometimes many of them emerge at once. All these drops of reddish material look like a red rain dropping from roofs and trees. Superstitious people used to call this a shower of blood. They often thought it meant the end of the world. Of course, it was really the beginning of new life for the butterflies.

When finally the great day arrives, the industrious moth, which went into its closet in its work clothes, is ready to emerge in its best party dress.

Jack Frost Arrives

The temperature of an insect changes according to the weather in the insect's surroundings, so its temperature is always changing.

In the winter, the insects hidden in the ground and under the bark of trees are just about as cold as the snow. They are so cold that they can hardly move at all. In the summer, they are nearly as hot as the sunshine. Then they run and fly very quickly.

If you listen to a cricket chirping, you can judge the temperature outdoors. The warmer the day, the faster the song. The snowy tree cricket sings the same musical note over and over. Count the number of times it sings in fifteen seconds, then add forty. The resulting number will approximate the reading on a thermometer.

The fur of a monkey is a warm hiding place for insects.

You might think that because insects get so cold in winter, you would want to bring them all in by the fire where it is warm. But if they were kept alive and active in a warm house, they would starve without any food. So it is better that they spend the winter months outdoors. There they just remain quiet until spring comes again.

Sometimes insects come out on a warm day. Then you see flies buzzing around the sunny side of a house. Sometimes you see caterpillars crawling slowly on the bark of trees. The mourning cloak butterfly often comes out on a sunny January day. It looks quite out of place sailing over the patches of snow.

Insects may spend the winter as a pupa or larva. Other insects lay their eggs during late summer and then die. The only thing that keeps the species from dying out completely is the cluster of eggs. Like tiny seeds, they wait for spring. Then, sure enough, they hatch and grow up to be just like the parents they never saw.

A few insects are active all year, even where the winters are cold. Lice and fleas that live on birds and animals keep warm in the thick fur and feathers. Cave insects crawl around as usual, for the temperature hardly changes at all in a cave.

Even on a winter day when the temperature is far below zero, and snow and ice are everywhere, bees are active in their hives. If you should visit an apiary (a place where bees are kept) on a winter day, put your ear against a beehive and listen. You will hear a faint humming sound. In the hive, even on a cold day, bees move around slowly, buzzing their wings. This activity keeps them warm enough so that they don't freeze.

A warm beehive sometimes attracts mice and other animals. If a mouse finds the hive, it may eat some of the honey the bees have stored for food. It may build its nest in front of the entrance so that the bees cannot get out in the spring.

Often the bees drive the mouse away with their stings. Sometimes they sting it so much that it dies. Then they have to leave the body there. But the bees often cover a dead mouse with their wax, sealing it up so that the air in the hive will stay fresh.

A few insects go south in the winter, just as the birds do. The large orange-and-black monarch butterfly may travel from Canada to Mexico. It goes in flocks of thousands. Sometimes it crosses many miles of water over the Great Lakes and the Gulf of Mexico. Nobody yet knows how it finds its way. It is one of the greatest of all insect travelers.

The Insect Parade

Do you know the difference between a fly and a bee? Can you tell a moth from a butterfly? Are termites really white ants? You can have fun learning to tell insects apart.

You don't have to live in the country to find plenty of insects. You can find them easily enough in the city, too. Several different kinds may sometimes be found in a city schoolroom. They come in through the open windows on a warm day and usually collect on the window panes.

There are about two dozen different groups of insects. Each group is called an order. The common insects belong to about eight orders. The following discussion will help you to know more about the kinds of insects you find and the groups to which they belong.

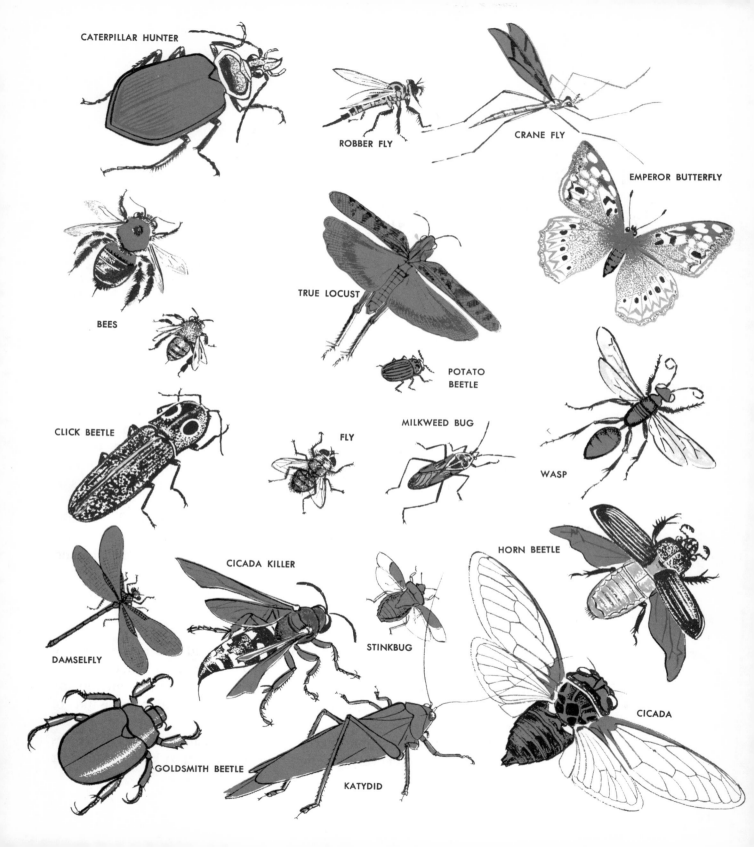

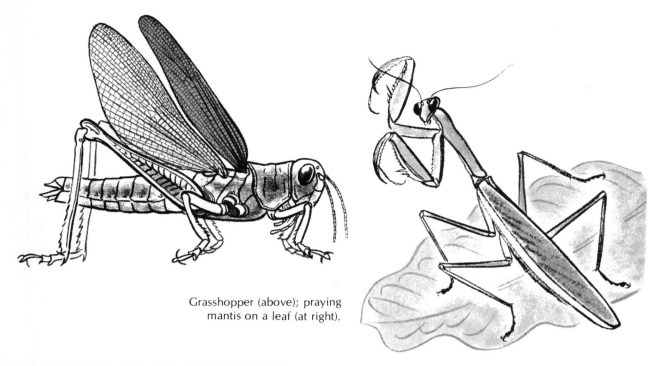

Grasshopper (above); praying mantis on a leaf (at right).

1. RELATIVES OF GRASSHOPPERS

The praying mantis is the terror of the insect world. It catches and eats nearly every kind of insect it can find. It belongs in this group.

The walking stick is also a relative of the grasshoppers, and so is the cockroach. Walking sticks eat plant leaves and twigs, but cockroaches eat nearly anything. Cockroaches have even eaten the glue from the backs of postage stamps.

Crickets and grasshoppers are the most musical of insects. They make most of the insect sounds you hear in the country. The mole cricket even "sings" under the ground. Locusts buzz their wings together as they fly, and katydids "call" from the treetops at night.

2. THE DRAGONFLIES

Walk near a swamp or brook in summer, and you may see many dragonflies. People used to think they would sew up your ears when you were asleep. They called them "darning needles." Of course, they don't do any such thing. They are really helpful insects, for they catch thousands of mosquitoes.

If you catch a dragonfly, notice its large eyes and odd legs. The eyes help it to see in almost every direction. The legs form a basket to catch other insects as it flies along. With its wings pointed out to the sides, it looks like a tiny airplane.

The nymphs of some dragonflies can travel by "jet propulsion." They squirt water out of the end of their bodies, which makes them shoot forward like a little jet airplane. Perhaps it's more like a submarine, though, for they live underwater.

Damselflies look like dragonflies and belong to this group. They fold their wings and point them up in the air.

3. THE FLIES

When you catch a fly or mosquito, count its wings. The total may surprise you. All the other common insects have four wings, but the flies have only two. Instead of a second pair, they have a pair of knobs attached to the thorax. If these knobs are hurt, they cannot fly.

One of the strangest flies lives in the fur of some animals. It has no wings, and it runs around in the hair of sheep, goats, and deer. It looks like a large flea. A few other wingless flies live in the feathers of some birds. Still another, a wingless crane fly, can sometimes be seen walking around on snow. It looks like a spider but has only six legs instead of the spider's eight. It is one of the first insects to come out in the spring.

A few tropical flies are among the most dangerous of all insects. The anopheles mosquito carries malaria disease from one person to another. The aedes mosquito carries yellow fever. The tsetse flies of Africa carry sleeping sickness. Houseflies may go right from a garbage pail to your dinner table. Doctors worked many years before they found ways to control these insects. Many of them caught the same diseases they were fighting.

Some flies look like other insects. Some are colored exactly like a bee. Others look like wasps or hornets. Some look like moths. But if you count the wings, you'll see that they are not wasps or bees at all—they are flies.

The bite of the black widow spider is very poisonous. But one tiny fly

Bee fly.

Tsetse fly.

Anopheles mosquito.

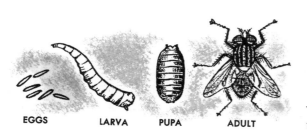
EGGS   LARVA   PUPA   ADULT

The life cycle of the housefly includes four stages.

has learned to go right into this spider's web without getting caught. It lays its eggs on the spider's egg sac, and the maggots burrow inside. They eat the eggs of the spider. Then they fly away. Without these little flies there would probably be many more black widow spiders.

4. THE BEETLES

There are more kinds of beetles than any other insect group in the world. If you began collecting beetles at the rate of one new kind every day, your life wouldn't be long enough to collect all of them. It would take several hundred years. There are more than 250,000 known kinds.

There are some beetles so small that they could hide beneath a single grain of salt. On the other hand, there are also some beetles so large that when they spread their legs they would cover half this page.

Perhaps you remember the Bible story of David and Goliath. One of the largest of all insects is the Goliath beetle. It has a body almost as long as a banana. In fact, when a living specimen was sent to a museum, scientists found that it liked to eat bananas. It was found in Africa and was quite lucky to stay alive long enough to reach America. Some natives in Africa like to fry the giant creatures in oil for food.

Far left, close-up of the head of a male stag beetle. Left, stag beetles, male (top) and female (bottom).

You can usually identify a beetle easily when you see it. It has powerful jaws for chewing, and its heavy wings look like two shields on its back. Underneath are the folded wings that are used in flying. The wings of the Goliath beetle may spread eight inches.

Stag beetles have jaws so large they look like the antlers of a tiny deer. Ground beetles have powerful jaws for eating other insects. The jaws of the boll weevil are at the end of a long snout. It looks like a true bug at first, but if you look closely you'll see its jaws.

5. THE TRUE BUGS

Not all insects are bugs. The only real bugs are those with the drinking-straw mouths made for puncturing plants or drinking the blood of animals. Bugs have four wings or no wings at all. Half of the wing is tough, like a beetle's. The other half is thin, like that of a fly.

Squash bugs, bedbugs, and stinkbugs are all true bugs. So is the diving water boatman with its long legs that look like oars. Ladybugs and June bugs are not really bugs at all. They are beettles with chewing mouths.

The noisy cicada is a relative of the bugs. So are the green aphids.

A giant water bug makes a meal of a tadpole. The insert shows a close-up view of the head and mouth of a true bug.

Aphids are interesting because they give off a sweet fluid called honeydew, which ants like to eat. Some ants even carry aphids down into the ground to feed on the roots of plants. Then they have a honeydew supply right in the nest. This is almost like a farmer who keeps cows for milk.

6. THE MOTHS AND BUTTERFLIES

Rub your finger gently on a wing of a butterfly or moth. You will find that a soft powder comes off. A microscope would show you that this powder is really thousands of tiny scales. They are arranged on the wing like shingles on a roof.

Some moths are not much larger than a pinhead. The largest may have a wingspan of more than a foot. Some moths are the most colorful of all insects. They may shine bright blue in one light, green or purple in another.

Moths and butterflies have a coiled tube for sipping liquids instead of the pointed beak of the bugs or the jaws of the beetles. They poke this tube down into flowers to get the sweet nectar.

One time several butterflies, near a group of soldiers standing at attention, flew from one soldier to another. They alighted on each brightcolored shoulder patch and uncoiled their long tubes. The soldiers must have looked like some new flower to the butterflies.

Some moths are helpful to man. For example, we unwind silk from the silkworm's cocoon. Many caterpillars eat troublesome weeds. But many butterflies and moths have young that are not so helpful. They eat our gardens, our clothes, and our forests. Gypsy moth caterpillars may eat the leaves from hundreds of trees at once.

Are there any differences between moths and butterflies? Look at the way a moth holds its wings. The wings lie down flat over the moth's sides and back. A butterfly holds its wings pointed up over its back. Moths have antennae that look like feathers. The antennae of a butterfly look like long threads with a knot at the end. And, of course, you usually see moths at night and butterflies during the day.

If you want to learn to tell a butterfly from a moth, look closely at the wing scales, the head, tubes, and antennae. Also study the different position of the wings when resting. (See illustrations on opposite page.)

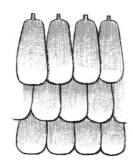

On the left are some enlarged wing scales of a butterfly. Note the rounded edges.

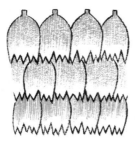

On the right are some moth scales. Their edges are much more ragged.

Close-up of head and coiled tube of moth.

Close-up of head and tube of a butterfly.

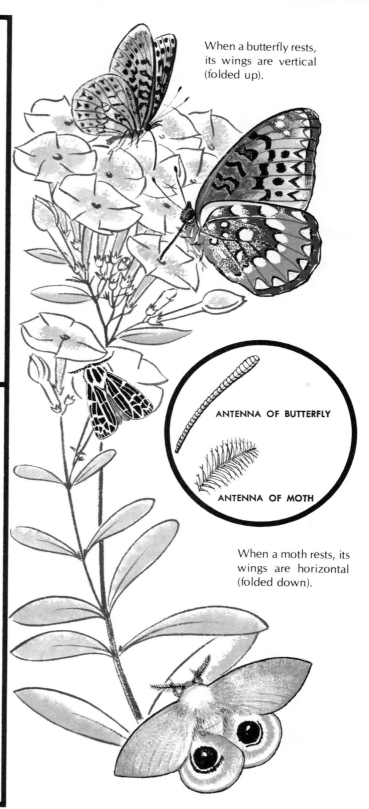

When a butterfly rests, its wings are vertical (folded up).

ANTENNA OF BUTTERFLY

ANTENNA OF MOTH

When a moth rests, its wings are horizontal (folded down).

47

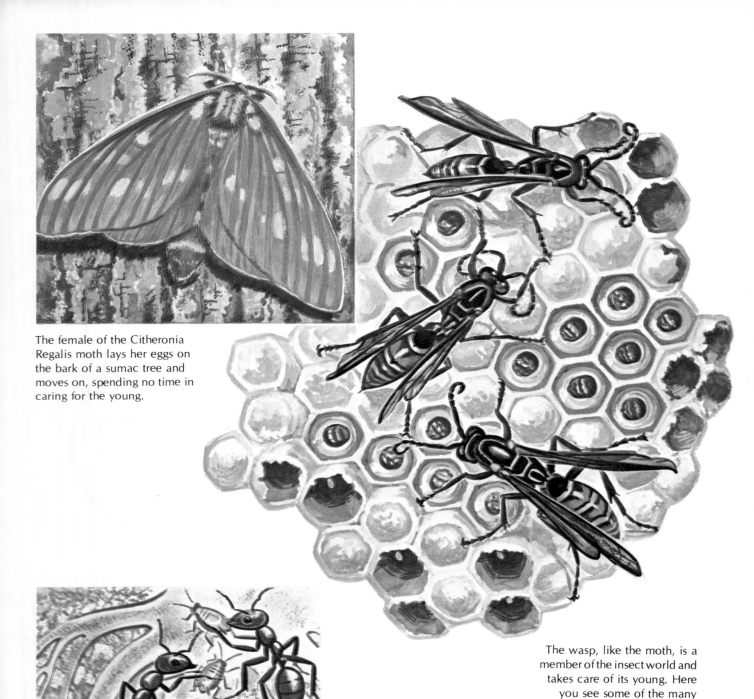

The female of the Citheronia Regalis moth lays her eggs on the bark of a sumac tree and moves on, spending no time in caring for the young.

The wasp, like the moth, is a member of the insect world and takes care of its young. Here you see some of the many workers feeding the babies in the wasp's nest.

Cornfield ants "milking" aphids.

ORDERS 7 AND 8—THE "SOCIAL INSECTS"

The female moth creeps slowly along a tree limb. Finally she stops and lays a few eggs, which stick to the bark. Then she crawls away.

When her babies hatch, they must make their way by themselves. There are no parents to protect them. Alone, they must find food and hide from their enemies.

Most other insects do the same thing. The mosquito sets her eggs afloat on a little raft. Walking stick insects just drop theirs on the ground. Other insects place their eggs where the young may find food and shelter, then leave them forever.

How strange it is, then, to see a wasp caring for her babies. She brings food, chews it until it is soft, and places it in their little mouths. She licks them, strokes them with her antennae, and builds a shelter to keep out the sun and the rain. If danger strikes, she flies toward it even if it is an animal a thousand times her size.

A bee will defend the hive against a larger enemy (below left). Ants, like bees, are social insects. The fire ants below carry a larva and a cocoon to a safer place. In the bottom illustration you see two termites licking each other. This is a process that, according to scientists, helps to keep the insect family together.

Of all the insects, only ants, termites, wasps, and bees take care of their families. The family, in turn, helps the mother to care for her later offspring. Because of this, and also because these insects live in little groups or societies, they are called "social insects." There are about ten thousand kinds of bees known in the world today. Only a few of them are social kinds, living in families, as for instance the honeybee.

Nobody is sure how the social habit developed, but scientists have noticed an interesting thing. Often, as soon as the young insects are given food, they produce little bubbles of saliva. The adult insects lick up the saliva quickly and then produce more saliva, which other adult insects lick up.

Does this special substance in the bubbles help to keep the insect family together? Many scientists think so. They call the process trophallaxis, or "nursing together." As one scientist said, "To others in the nest, each insect must be a living lollipop."

7. ANTS, BEES, AND WASPS

If you see an insect with a slender waist, the chances are that it is an ant, a bee, or a wasp. If it has four transparent wings, you can be almost certain of it. Some flies and moths look like them, but flies have only two wings and moths have thick bodies.

Although there are species of wasps and bees that do not live in

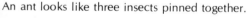
An ant looks like three insects pinned together.

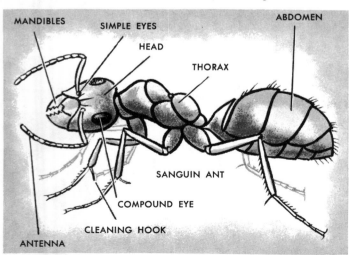

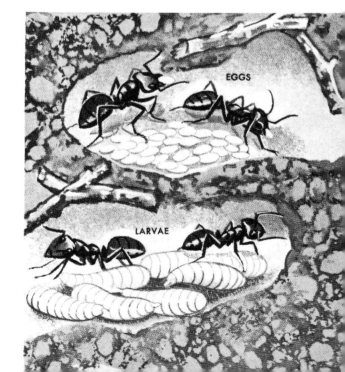

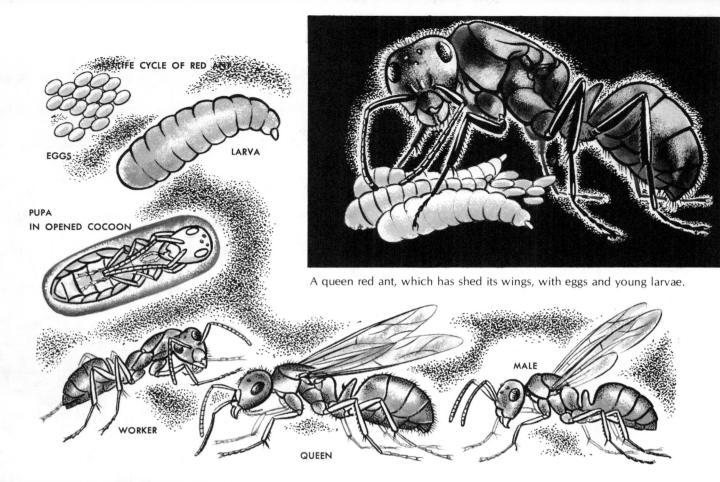

A queen red ant, which has shed its wings, with eggs and young larvae.

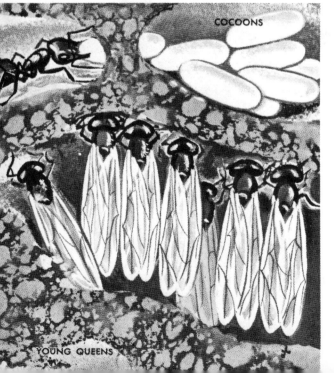

Ants usually keep their young of different ages in different rooms. Here is a cross section of a formica ants' nest with a group of winged young queens, eggs, cocoons, and larvae, all separated and cared for by workers.

societies, there are no solitary ants. Even the most primitive kinds are organized into communities, and the ants are the most highly developed species of the insect world.

Like the bees and wasps that live in communities, the ants have classes, or castes, among the adults. Like bees, the workers are females, mostly unable to lay eggs, but their occupations are more numerous than those in

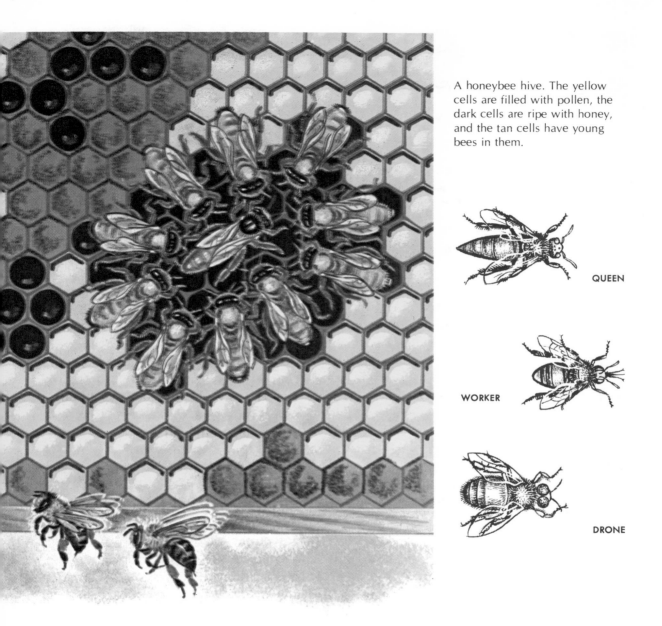

A honeybee hive. The yellow cells are filled with pollen, the dark cells are ripe with honey, and the tan cells have young bees in them.

QUEEN

WORKER

DRONE

the bee society. The worker is usually much smaller than the queen and has no wings. Ant queens and males have wings that they use in their mating flight and while searching for new colony sites.

After having mated in the air, the male flutters dying to the ground. The queen glides to a landing, then does an astonishing thing. Hooking her legs up over her shiny wings, she twists the wings until they break off.

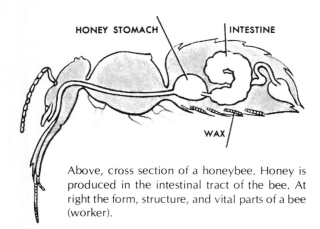
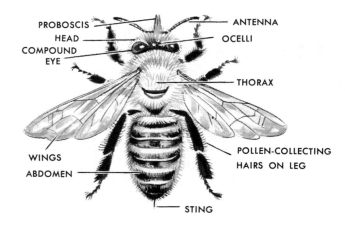

Above, cross section of a honeybee. Honey is produced in the intestinal tract of the bee. At right the form, structure, and vital parts of a bee (worker).

Only the bumblebee, which can reach the nectar, is able to pollinate red clover.

One after the other, the wings fall off. Now she looks strange and humpbacked. Her huge muscles no longer have wings to operate, but they will soon have another job to do.

She hunts quickly for a place to hide. Burrowing under a stone, she digs into the earth. There she makes a little chamber and seals it tight. Alone in the dark, she begins her new family.

Here her great wing muscles are put to work again. They begin to shrink, giving nourishment back to her body. As she absorbs their strength, she lays half a dozen eggs. In a day or two a few more are produced. Finally she has a little cluster of eggs, the beginning of a new anthill colony.

Even in early history honey was obtained from wild and domesticated colonies of bees. Therefore the honeybee is probably one of the best-known insects. An average colony, consisting of a queen, workers, and drones, can contain forty thousand to fifty thousand bees.

Regularly bees swarm, which means that the old queen flies off with thousands of her "subjects" to start a new colony, leaving behind a young queen with the rest of the hive's population.

Each one of the thousands of bees in the hive has a job to do. There are no loafers. There are no supervisors to see that all are working. Every bee does its work without being told. There are enough workers of each kind—nurses, honey fanners, pollen gatherers, and wax makers. If the hive gets too hot, some of the bees set up an air current with their wings as a ventilation system. If more wax is needed, extra bees set about making it.

Nurse bees are walking all over the nursery or brood comb. Sometimes they put their feet right on the heads of the babies. One after another, they bend down and poke their heads inside the cells. They feed and lick their little sisters in their cradles.

There is another kind of cell on the edge of the comb. It is much larger than the others and looks almost like a peanut shell made out of wax. Inside it is a larva just like the thousands of others—only a little larger. This strange cell must be something special to be off on the side of the comb where it can have plenty of room. Indeed it is, for it is the royal nursery of the larva that will soon become the new queen.

One kind of bee emerges from the "nursery" that looks different from the others. Its head seems to be completely covered with two huge eyes, like a helmet. Its legs are shaped differently; its body is longer. It walks up to a worker bee and seems to ask for food. The worker stops work and gives it a bit of honey. Then it goes to another for more food, and so on. How strange to see this kind of bee in a bustling hive where all the others are working!

Yet this bee has a job, too. It is one of the most important of the bees. It is a drone, or male bee, one of a few dozen brothers among the thousands of worker sisters. Its job will begin when that peanut-shaped queen cell has opened and the new queen has made her way into the world. For the drone must provide millions of tiny sperm cells that the queen bee stores in a special pouch in her body. Then, just before she places new eggs in the brood comb, she fertilizes each one with a sperm cell so that it can develop into a new worker.

CARPENTER BEE

MASON BEE

Once it was thought that the drones were lazy because they didn't help with the hive duties. But now we know that they couldn't help, no matter how much they wanted to. Their legs, heads, and bodies are not shaped right for fashioning the wax into perfect little chambers. They don't even have a sting with which they can help to drive away enemies. They just have those important little sperm cells, ready to be given to the queen in mating.

Not all bees live in hives. Carpenter bees dig holes in wood. Bumblebees make their home in holes in the ground. Sometimes they use an old mouse nest. They seem almost to be paying the mice back for living in the hives of honeybees.

Mason bees lay their eggs in an old snail shell or a knothole. A mason bee might even use a keyhole in a door for its home, cementing it shut with sand and clay.

Nearly everybody likes fig bars. There wouldn't be a single one if it weren't for the tiny fig wasp. She wanders around on the blossom of the fig, spreading the pollen and making it possible for the fruit to develop. Her young hatch out and seek new blossoms. This valuable insect was actually imported into California for the production of Smyrna figs.

There are about ten thousand kinds of wasps. Only a few of them are social. Most of them live their lives quietly and unseen. Some of their

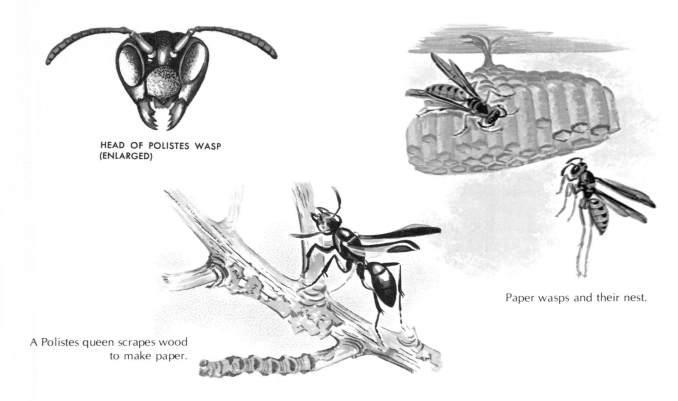

HEAD OF POLISTES WASP (ENLARGED)

Paper wasps and their nest.

A Polistes queen scrapes wood to make paper.

relatives, the horntails and sawflies, attack trees and fruit, but a great many of the true wasps are as useful as honeybees.

Some wasps hunt and kill spiders. Others catch harmful caterpillars. Some of the smallest wasps thrust their eggs into the bodies of our garden pests. Then the tiny babies bore through the insect and kill it. They may be no larger than the size of the period at the end of this sentence.

How long ago was the first paper made? Was it over five hundred years ago, when Johann Gutenberg printed a Bible in 1455? Two thousand years ago, when the Chinese made paper from mulberry bark? Four thousand years ago, when the Egyptians wrote on flat sheets from the papyrus reed? The truth is that paper was first made *millions* of years ago. No human hand touched that first sheet of paper. It was probably manufactured in much the same way it is today—by a queen wasp, making a shelter for her eggs and larvae.

She goes to a piece of dried wood on a tree or the side of a house. Her strong little jaws work like scissors. She cuts chunks of wood and mixes them with her saliva. Getting all she can carry, she sometimes tucks extra pieces "under her chin" between her head and her first pair of legs. Then she flies back to her home, chewing on the pieces. When she spreads out the

YELLOW JACKET

mixture, it dries to a tough paper. This material she shapes into her "paper palace," nests you can see illustrated on this and the opposite page.

It is a mystery how she determines where she will place her new home. Often it is on the end of a tree branch or under the roof of a house. But sometimes she chooses strange places. One wasp nest was built high on a factory chimney, next to the whistle. Every time the whistle blew, hundreds of wasps buzzed in all directions.

Just like the bees, wasps are divided into workers, drones, and queens. Most wasp paper palaces last only a few months; on the other hand, a beehive may last for years.

The queen chews the wood until it makes a sticky paste. Then she plasters it on the underside of a branch or roof or the interior of an animal's den. Even a family of skunks has to move out when the wasps move in.

The queen makes a little six-sided paper cell that is shallow at first and hangs downward. She places a single egg in the cell, covered with a sticky material so that it won't fall out. A few more cells and eggs complete her little nest.

What is the difference between a wasp and a hornet? Hornets are short, active wasps that live by the hundreds in their paper nests. Yellowjackets and white-faced hornets are actually special kinds of wasps.

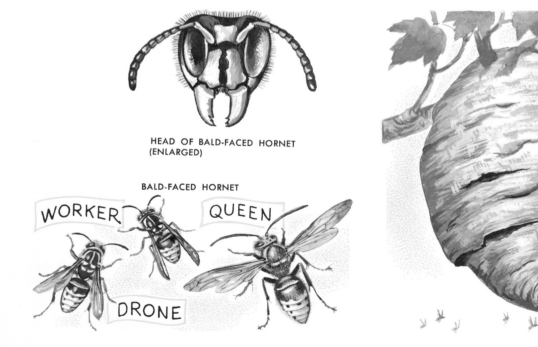

Hornets' nest with some adult insects.

HEAD OF BALD-FACED HORNET
(ENLARGED)

BALD-FACED HORNET

WORKER   QUEEN   DRONE

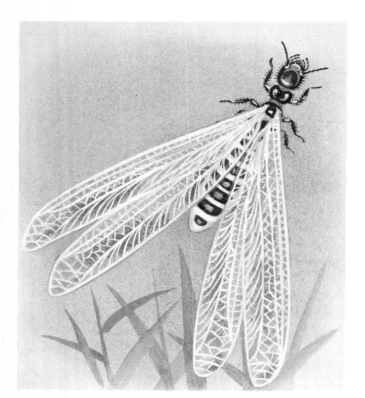
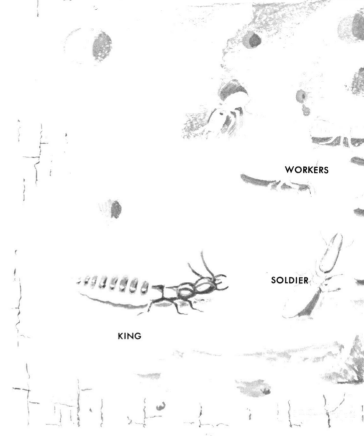

8. THE TERMITES

If you live in Canada or the northern United States, you may not have seen many termites. They are much more common farther south.

Sometimes termites are called "white ants," but they are really not ants at all. Ants have a thin waist bewteen the thorax and abdomen. Termites are thick-bodied from head to tail.

Soldier termites guard the nest from enemies with their powerful jaws. Hundreds of workers build the nest and get the food. The queen lays great numbers of eggs. Sometimes there may be more than one queen, and a few kings as well.

Sometimes you see termites by the thousands as they come out on a windowsill or an old stump. These are dark-colored kings and queens ready to leave the nest. They fly to a new spot and then do a strange thing. They break off their wings so they can never fly again. Then they burrow into the ground and start a new colony.

EGGS

QUEEN

The termite "majesties" in the "throne room" of their "castle" (a chewed-up wooden house). The queen is fed by the workers, another workers cares for the eggs, and a soldier and the king stand by. These are wood termites.

The wings of the termite queen on page 58 look as if they were made of very delicate lace.

Compare the illustration below with the diagram of the ant on page 50. It shows why the termite, which is often called the white ant, is as far removed from the white ant as a horse is from a hippopotamus. Their bodies are built differently, the larvae develop differently, and the organization of the societies is different. A termite is not a white ant.

TWO TERMITES
(ACTUAL SIZE)

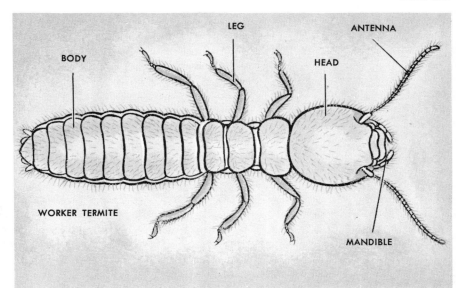

WORKER TERMITE — BODY, LEG, HEAD, ANTENNA, MANDIBLE

Termites eat nearly everything made of wood, leaving only a thin outer layer. Once a teacher opened an old desk drawer. Termites had drilled up through the floor and into the desk leg. They had hollowed out the wood of the desk until it was just a shell. When he pulled on the drawer, the desk toppled.

Insects and Plants

Many kinds of insects cause galls on plants. A gall is a swelling or bump. Some are made by flies. The mother fly lays her eggs in the stem of a plant. The stem begins to swell. The eggs hatch inside the swelling. Then the maggots live in their strange house.

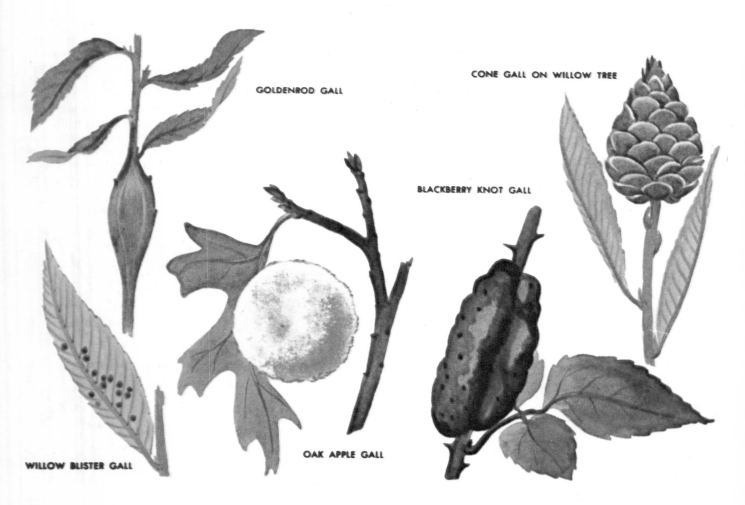

 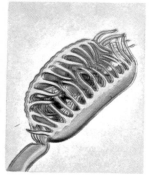 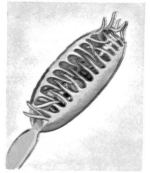 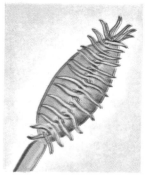

The Venus's-flytrap catches an insect.

An insect stuck in the hairs of a sundew plant.

A few plants feed on insects. They are called insectivorous plants. The pitcher plant has leaves that are hollow and shaped like a flower vase. Rain water falls into them and makes a little puddle. Insects fall into the water and drown. Then the plant digests the insects, somewhat as you digest the food you eat. Some pitcher plants are so large that they may trap frogs, lizards, or even mice.

The sundew has sticky hairs on its leaves. Insects land on the leaves and get tangled in the sticky surface. Soon, like those trapped by the pitcher plant, they are digested.

The milkweed catches insects, but it lets them go again. It has flowers with little traps in them. When an insect puts its foot in the trap, it is held fast. Then it struggles to get free. Finally the trap breaks off, and the insect flies away with it. Then when it visits another milkweed, grains of pollen in the trap fall out on the new blossom, so the milkweed can make its seeds.

Some plants even live inside the bodies of insects. One fungus attacks houseflies and kills them. Another kind attacks caterpillars. Bacteria, which are so small that you need a microscope to see them, kill many others.

Without these little plants, there would be even more insects in the world than there are now.

Many insects are useful in carrying seeds of plants. Small hooks on the seeds may catch in the hairs on the body of a fly or bee. Then the seed is carried through the air as the insect flies away. Later it drops off and starts a new plant. Some insects take seeds to their nests in the ground. The seeds may grow, starting a new plant right in the middle of the nest.

One of the most interesting seeds is the Mexican jumping bean. This is a seed that contains a small caterpillar. This little insect chews away at the inside of the seed. It changes position every few minutes. Every time it moves, the seed rolls around, just as you can roll a big box by moving around inside it. Finally the caterpillar turns into a little moth. Then it flies away to lay its eggs in new seeds.

Collecting Insects

You can make an insect collection of your own, which is one of the best ways to get to know the insects. You will need these things:

(1) A magnifying glass.
(2) A pair of tweezers.
(3) A few dozen pins. (Regular insect pins are best. Perhaps a biology teacher can help you get some. If not, you may have to use straight pins.)
(4) A box with a tight cover, such as a cigar box or a candy box.
(5) A piece of thick cardboard, cut to fit exactly into the bottom of the box. (With this, pins may be stuck in easily.)
(6) A killing jar. (This should have a tight lid. A pint-size jar will be fine. Put a crumpled piece of paper towel on the bottom, wet it with a few drops of cleaning fluid, and force a circle of cardboard into the jar a little above the paper so that the insect cannot touch the damp paper. Keep the jar tightly closed when you are not using it.)

After you catch an insect, put it in the bottle for five minutes. It will quiet down right away. When it is still, take it out with the tweezers.

To keep your insect in good condition, carefully stick a pin through its thorax or chest from the top. Push the pin down until the pinhead is about one quarter of an inch above the insect's back. Beetles should have the pin stuck through the right wing. Put a small label on the pin, telling where and when you found the insect. If you know its name, put this on another label.

Stick the pin into the soft cardboard bottom of the box, and you'll be able to look at the insect whenever you wish. Always handle it with care after it is dry.

You can mount tiny insects, too. Glue them to one corner of a three-cornered piece of paper. Then push your pin through the center of the paper.

Butterflies and moths should have their wings spread. Do this as soon as possible after collecting them. Don't let them dry out. Spread the wings flat on a piece of soft wood, one at a time, until all four wings are out straight. Hold them in place with strips of waxed paper. Never put pins through the wings.

If an insect gets hard and dry, it can be softened and relaxed with steam from an iron. Place it in a saucer and let the steam from the iron point right at the insect. Or put it on a piece of wire screen over hot water. In a few minutes you can handle the insect without breaking it.

Put a few moth crystals in the box every three months. These are the same crystals used to protect winter clothing when it is stored away in the spring. Then other insects won't get in and eat your collection. You can also keep insects in tiny bottles of alcohol. Regular rubbing alcohol is good. The colors will soon fade, but the insects will stay soft.

You can make a little insect aquarium. Fill a goldfish bowl half full with water. Put in a few pebbles and weeds for hiding places. Keep it out of bright light or the water will turn green. Then you can put any water insect in your aquarium. Cover the top, because most water insects can fly.

Your pets will feed on a bit of liver or fish. Serve it to them on a pair of tweezers, or hang it in the water by a thread. In an hour, take out all food that has not been eaten. In that way, decaying food will not foul the water.

You can make a fine display with your insect collection. Beetles, grasshoppers, and dragonflies can be mounted in special boxes and then hung on the wall of your room. Mounted butterflies and moths make interesting displays to hang on walls or to give as gifts.

To make a display case, find a large flat box, such as a writing paper box or candy box. Measure it and cut a piece of glass that will just fit over the box. Glue a picture hanger on the back of the box so it can be hung up later. Fill the box with cotton. It is a good idea to put some moth crystals in the cotton as protection against other insects.

Place your insects carefully on the cotton. Press them down so they will stay in place. Butterflies and moths should be mounted while they are still soft and flexible. Then put the glass plate over them and seal it neatly around

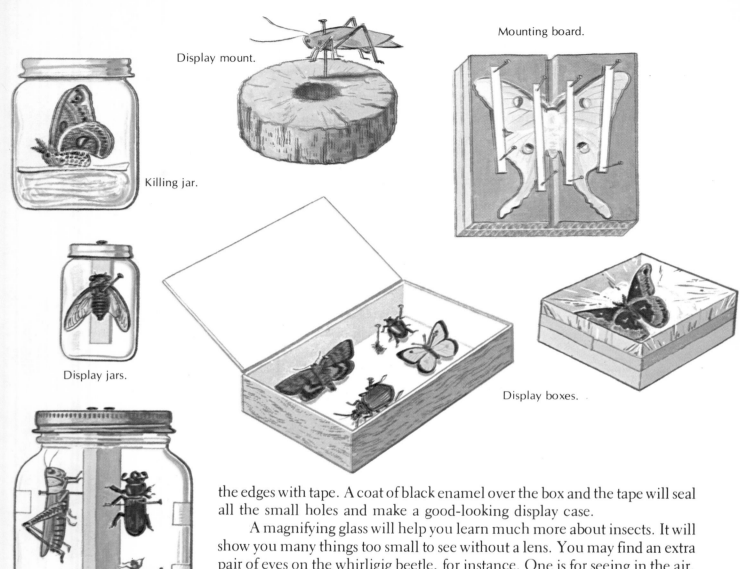

Display mount. Mounting board. Killing jar. Display jars. Display boxes.

the edges with tape. A coat of black enamel over the box and the tape will seal all the small holes and make a good-looking display case.

A magnifying glass will help you learn much more about insects. It will show you many things too small to see without a lens. You may find an extra pair of eyes on the whirligig beetle, for instance. One is for seeing in the air, while the other is for looking under water. If you look in the center of a daisy, you'll see tiny black thrips. A close look at their legs shows that they walk around on feet that look like little balloons.

In a few years, space ships may take men to Mars. They may even go beyond our own solar system. Some experts think they'll find strange new forms of life. But with a magnifying glass and a good sharp eye, we can stay home and find strange creatures of our own. Few animals that have ever lived are much stranger than the flies and beetles and bugs that live in the world right at our fingertips.